TOOLS FOR CAREGIVERS

- **ATOS:** 0.6
- **GRL:** C
- **WORD COUNT:** 34
- **CURRICULUM CONNECTIONS:** insects, nature

Skills to Teach

- **HIGH-FREQUENCY WORDS:** and, big, I, in, see, their, they
- **CONTENT WORDS:** away, back, chirp, eat, fly, grass, grasshoppers, help, jump, land, legs, rub, together, wings
- **PUNCTUATION:** exclamation points, periods
- **WORD STUDY:** compound word (grasshoppers); long /a/, spelled ay (away); long /e/, spelled ea (eat); long /e/, spelled ee (see); long /i/, spelled y (fly)
- **TEXT TYPE:** information report

Before Reading Activities

- Read the title and give a simple statement of the main idea.
- Have students "walk" though the book and talk about what they see in the pictures.
- Introduce new vocabulary by having students predict the first letter and locate the word in the text.
- Discuss any unfamiliar concepts that are in the text.

After Reading Activities

Have readers re-read the text on pages 12–13. Have they heard a grasshopper chirp before? It makes this noise by rubbing one of its big back legs on one of its wings. What other animals or insects make noise? Can the readers make the noises?

Tadpole Books are published by Jump!, 5357 Penn Avenue South, Minneapolis, MN 55419, www.jumplibrary.com

Copyright ©2020 Jump. International copyright reserved in all countries. No part of this book may be reproduced in any form without written permission from the publisher.

Editor: Jenna Trnka **Designer:** Michelle Sonnek

Photo Credits: IrinaK/Shutterstock, cover; Eric Isselee/Shutterstock, 1; FotoRequest/Shutterstock, 3; Grafissimo/iStock, 2bl, 4–5; Protasov AN/Shutterstock, 2tr, 2br, 6–7; pixelworlds/Shutterstock, 2ml, 8–9; Liudmila Gridina/Shutterstock, 2tl, 10–11; Biosphoto/SuperStock, 12–13; NHPA/SuperStock, 2mr, 14–15; Holger Kirk/Shutterstock, 16.

Library of Congress Cataloging-in-Publication Data
Names: Nilsen, Genevieve, author.
Title: I see grasshoppers / by Genevieve Nilsen.
Description: Tadpole books edition. | Minneapolis, MN: Jump!, Inc., (2020) | Series: Backyard bugs | Audience: Age 3–6. | Includes index.
Identifiers: LCCN 2018050520 (print) | LCCN 2018051537 (ebook) | ISBN 9781641288002 (ebook) | ISBN 9781641287982 (hardcover: alk. paper) | ISBN 9781641287999 (paperback)
Subjects: LCSH: Grasshoppers—Juvenile literature.
Classification: LCC QL508.A2 (ebook) | LCC QL508.A2 N55 2020 (print) | DDC 595.7/26—dc23
LC record available at https://lccn.loc.gov/2018050520

BACKYARD BUGS

I SEE GRASSHOPPERS

by Genevieve Nilsen

TABLE OF CONTENTS

Words to Know................................2

I See Grasshoppers............................3

Let's Review!................................16

Index..16

tadpole books

WORDS TO KNOW

eat

fly

grass

jump

legs

wings

I SEE GRASSHOPPERS

I see grasshoppers.

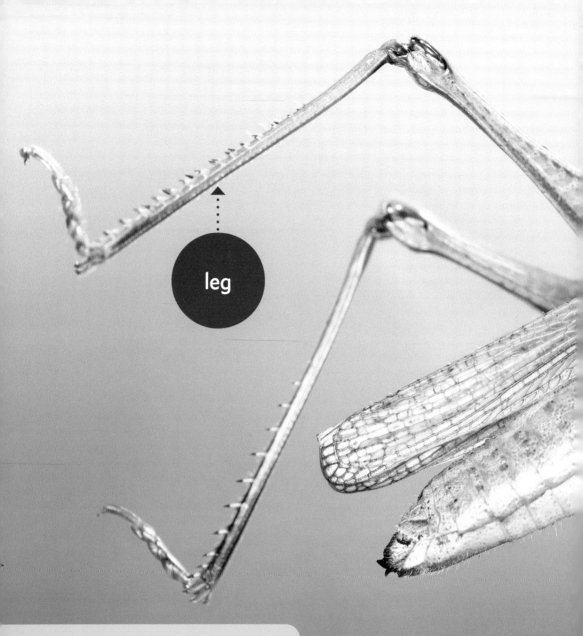

leg

They jump!

Big back legs help.

They fly.

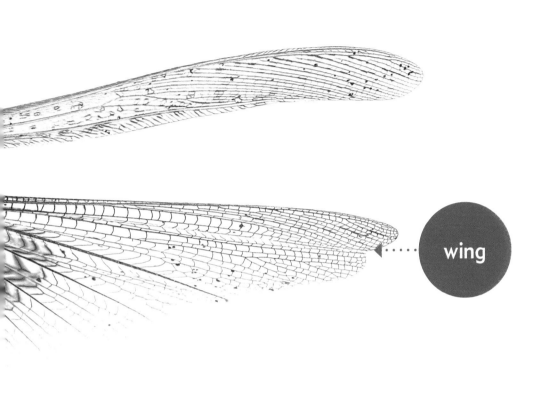

wing

Wings help.

They land in grass.

They eat.

They rub their legs and wings together.

Chirp! Chirp!